I0817148

Amazing Young People

PHILO FARNSWORTH

Martha London

DiscoverRoo
An Imprint of Pop!
popbooksonline.com

abdobooks.com

Published by Pop!, a division of ABDO, PO Box 398166, Minneapolis, Minnesota 55439. Copyright © 2020 by POP, LLC. International copyrights reserved in all countries. No part of this book may be reproduced in any form without written permission from the publisher. Pop!™ is a trademark and logo of POP, LLC.

Printed in the United States of America, North Mankato, Minnesota

052019
092019

THIS BOOK CONTAINS RECYCLED MATERIALS

Cover Photo: Underwood Archives/Archive Photos/Getty Images

Interior Photos: Underwood Archives/Archive Photos/Getty Images, 1; New York Public Library/Science Source, 5 (top), 20, 26; iStockphoto, 5 (bottom), 9, 11 (bottom), 12–13, 14, 22 (top), 22 (bottom), 27; World History Archive/Newscom, 6, 7; Time & Life Pictures/The LIFE Picture Collection/Getty Images, 8; Library of Congress, 11 (top); Red Line Editorial, 15; Bettmann/Getty Images, 17, 18, 21, 23 (bottom); Archive PL/Alamy, 19, 30; AP Images, 23 (top); Wikimedia Commons, 25; Matt Sullivan/The Journal Gazette/AP Images, 28; Shutterstock Images, 29, 31

Editor: Brienna Rossiter
Series Designer: Sarah Taplin

Library of Congress Control Number: 2018964785

Publisher's Cataloging-in-Publication Data

Names: London, Martha, author.

Title: Philo Farnsworth / by Martha London.

Description: Minneapolis, Minnesota : Pop!, 2020 | Series: Amazing young people | Includes online resources and index.

Identifiers: ISBN 9781532163685 (lib. bdg.) | ISBN 9781644940419 (pbk.) | ISBN 9781532165122 (ebook)

Subjects: LCSH: Farnsworth, Philo Taylor, 1906-1971--Juvenile literature. | Inventors--Biography--Juvenile literature. | Electrical engineers--Biography--Juvenile literature. | Television--Juvenile literature.

Classification: DDC 621.388 [B]--dc23

Pop open this book and you'll find QR codes loaded with information, so you can learn even more!

Scan this code* and others like it while you read, or visit the website below to make this book pop!

popbooksonline.com/philo-farnsworth

*Scanning QR codes requires a web-enabled smart device with a QR code reader app and a camera.

TABLE OF CONTENTS

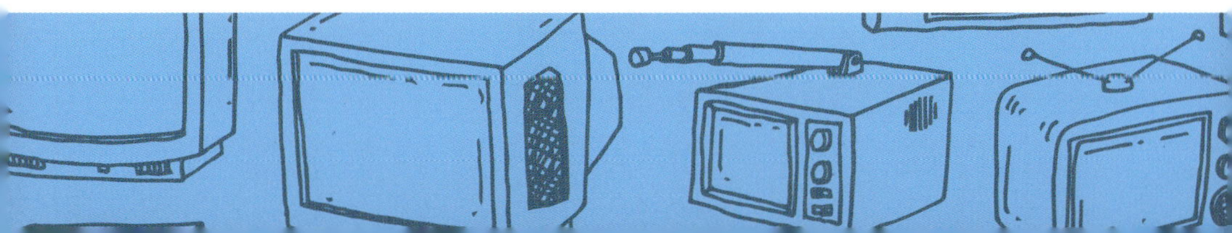

CHAPTER 1
IMAGES ON A SCREEN

Philo Farnsworth was an inventor. He designed the first electronic TV. Many inventors worked to create TVs in the early 1900s. People had been using radio waves to send sound through the air

since the late 1800s. But they wanted to send images too.

Philo's TV (top) was based on technology used by the radio (bottom).

Mechanical TVs used spinning discs to send signals.

Many early TVs used spinning discs. Each disc had several holes. A lamp shone through the holes to create and send images. This method was known as mechanical television. But sending images

this way was slow. And the discs could be quite large.

John Logie Baird showed the first image on a mechanical television in 1925.

John Logie Baird shows how an early TV works.

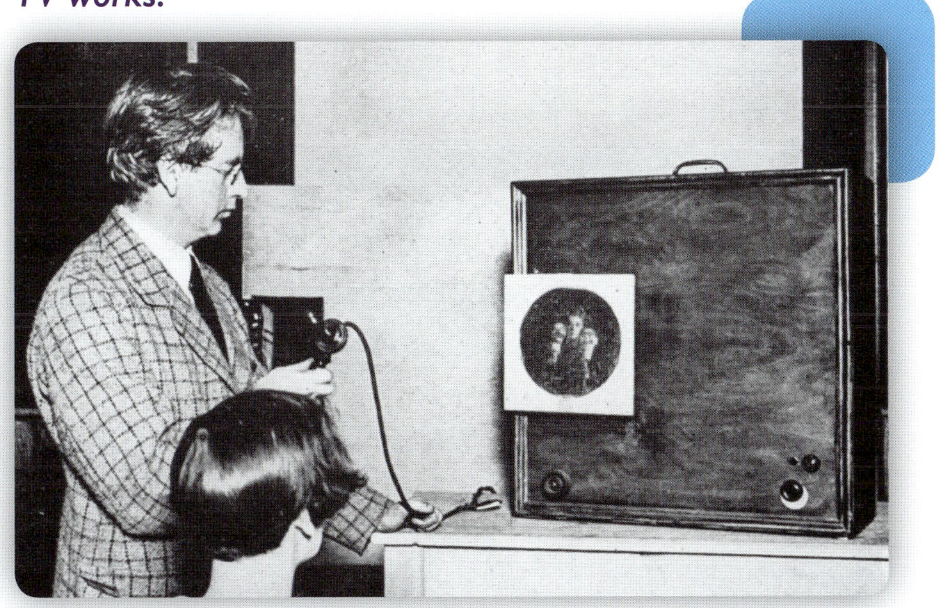

Philo holds a camera tube, his invention for sending images.

Philo's TV used electricity to send signals instead. His system did not need discs. As a result, it was faster and smaller. This new system helped TVs and **broadcasting** become common.

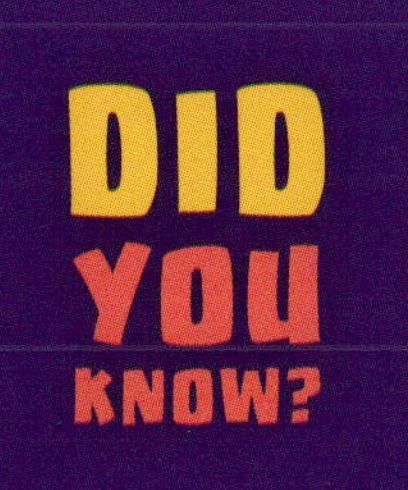

Each TV is part of a system. A camera records images. Other parts send the images out for TVs to pick up.

Cameras record the images shown on a TV's screen.

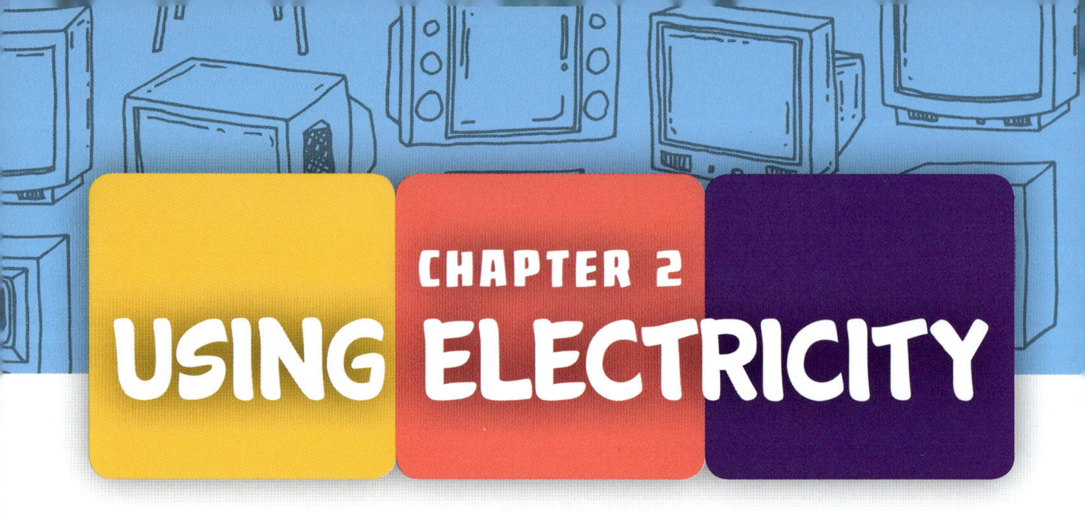

CHAPTER 2
USING ELECTRICITY

Philo Farnsworth was born on August 19, 1906. He grew up on a farm in Idaho. Philo loved learning. He read lots of books. He also loved figuring out how machines worked.

This photo shows a farm in Idaho in the early 1900s.

Philo loved fixing things, including his family's sewing machine.

One day in 1921, Philo was plowing a field. As he moved back and forth, the plow made lines in the dirt. The lines gave Philo an idea. The plow made one

Old-fashioned plows were pulled by horses.

line at a time. But it eventually covered the whole field. What if a TV could do something similar? The image could be broken up into tiny lines.

Philo made a sketch that showed how his idea would work. First, a camera would record an image. It would send each line of the image using electric signals. A **receiver** would pick up the signals. Going line by line, it would show the image on a screen.

Philo often showed sketches of his ideas to his chemistry teacher.

PHILO'S IDEA

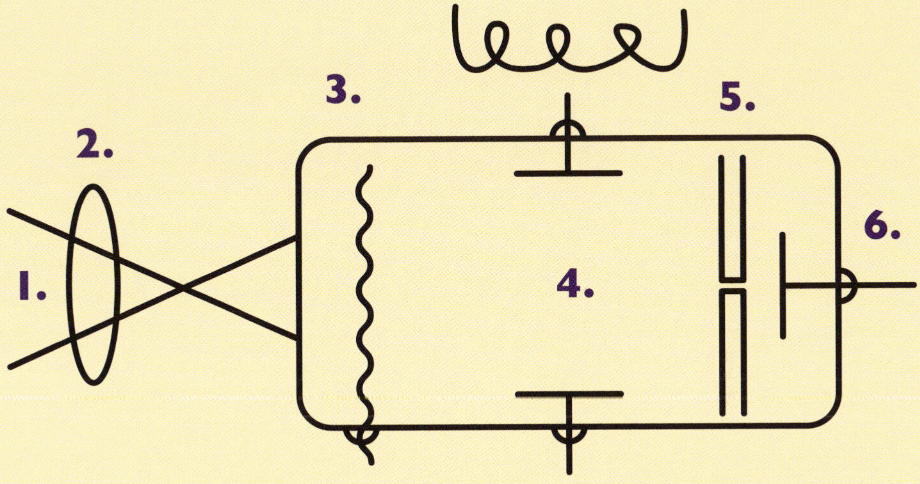

1. Light shines toward the camera tube.
2. A glass lens focuses the light.
3. The image shines through the glass to a plate.
4. A coating on the plate sends out tiny particles called electrons.
5. The electrons form a version of the original image.
6. A circuit sends out this image as electrical signals.

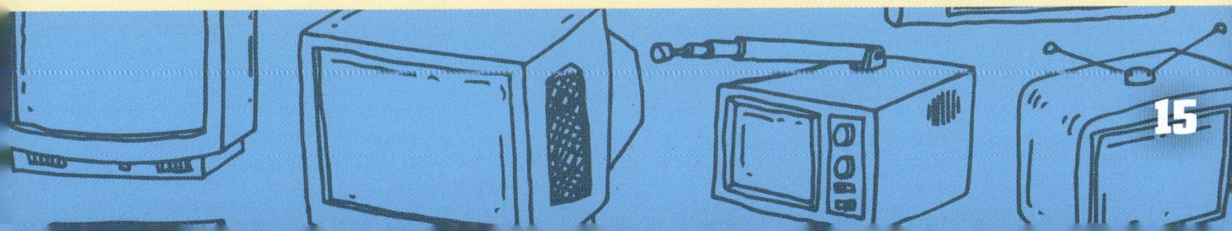

CHAPTER 3

TESTING AND IMPROVING

Philo spent the next several years working on his idea. He and his wife, Pem, moved to California in 1926. Together, they tried to build a working version of his TV.

Philo had to put many parts together to create a working TV.

PEM FARNSWORTH

Pem helped Philo create and test his TV. She drew each part of the TV. And she wrote notes about their work each day. Later, some people didn't believe Philo had invented the TV. They thought he was too young. But Pem's notes proved them wrong.

Philo created the first electronic TV when he was just 21 years old.

Pem and Philo stand next to a version of his TV.

On September 7, 1927, he sent the first electronic TV image. It was a single blurry line. But it proved that Philo's system worked. Philo received a **patent** for his system in 1930.

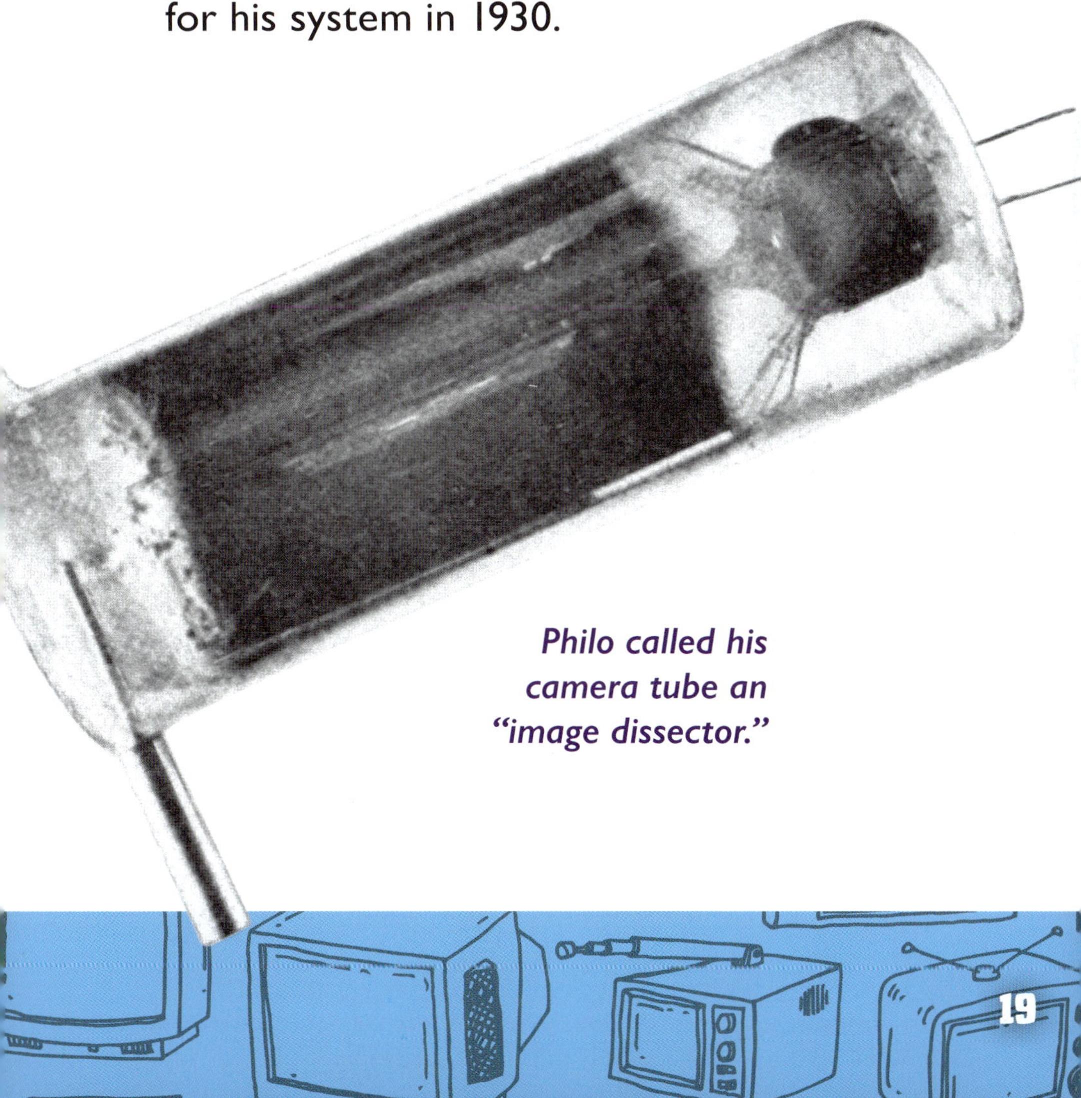

Philo called his camera tube an "image dissector."

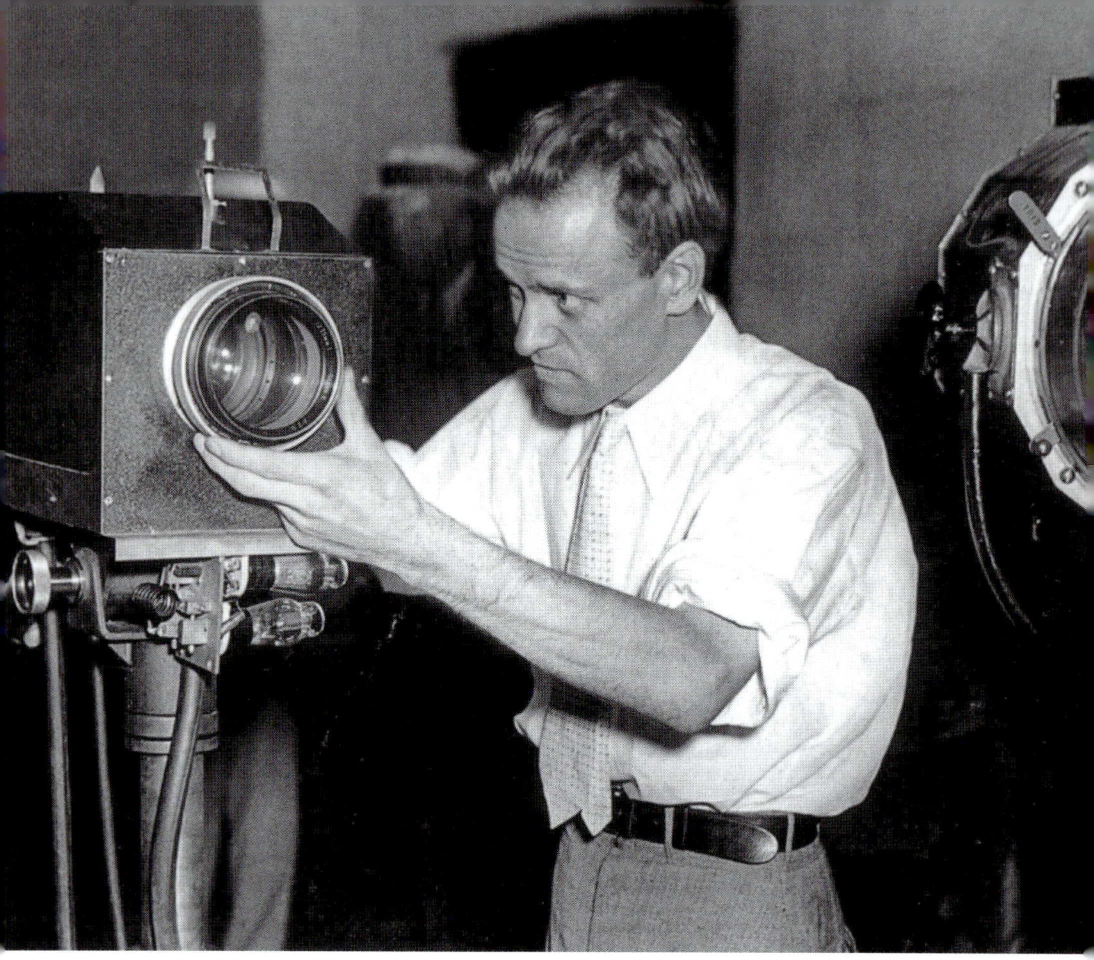

Philo adjusts a TV camera while demonstrating his system in 1934.

Philo hoped to make and sell TVs. He created new parts to improve his system. He also showed others how the

Philo's company films performers in 1935.

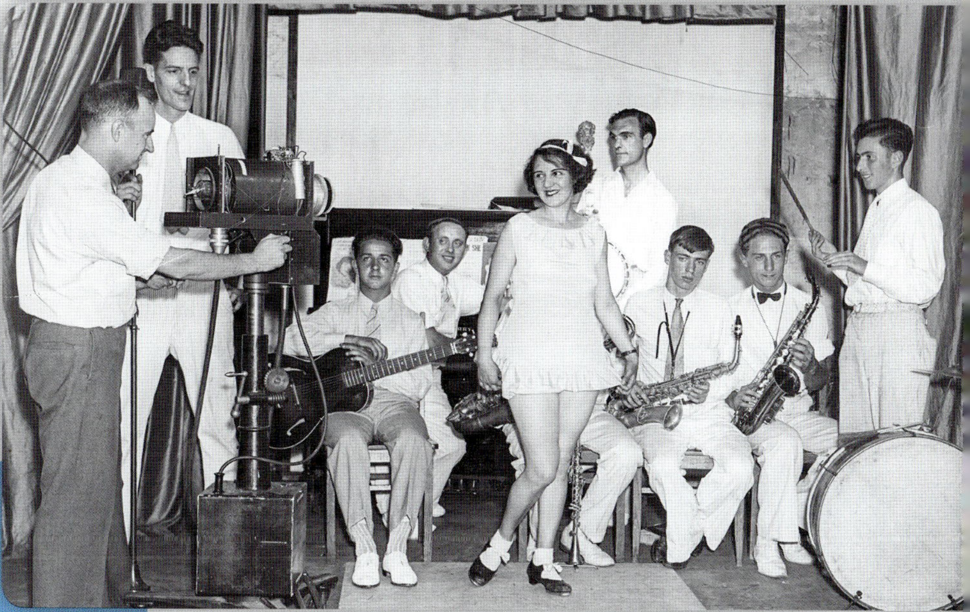

TV worked. He even formed a company in the 1930s.

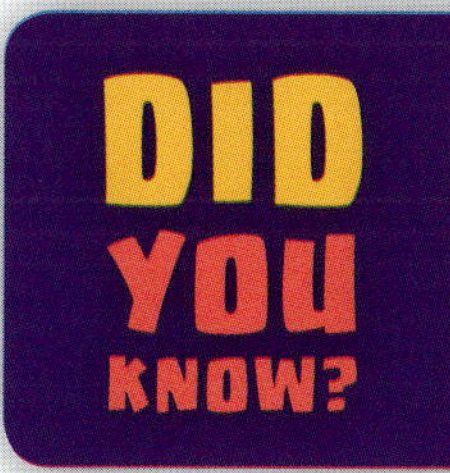

In total, Philo invented more than 100 parts for TVs.

1906

Philo Farnsworth is born in Utah on August 19.

1921

Philo draws the first design for an electronic television.

1926

Philo and Pem get married on May 27.

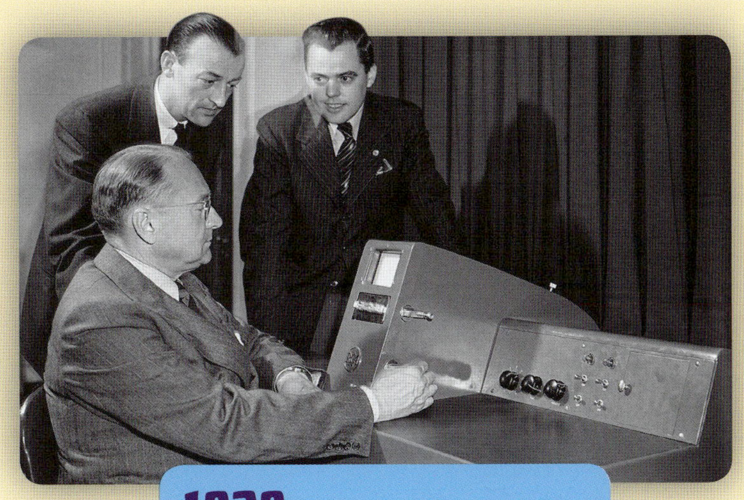

1939
RCA creates an electron microscope based on Philo's research.

1927
Philo sends the first electronic TV image on September 7.

1971
Philo dies on March 11.

CHAPTER 4

MANY INVENTIONS

Meanwhile, a huge company called RCA was also working to create TVs. They tried to take the rights for Philo's system. Philo fought back. In 1935, the government agreed the system was

Philo lived in this house in Fort Wayne, Indiana, from 1948 to 1967.

Philo's idea. Even so, Philo had to sell his company by 1949.

Philo spent many years researching nuclear energy.

Afterward, Philo worked on many other inventions. The **baby incubator** and **electron microscope** both exist because of him. He also researched **nuclear energy**. He hoped to find a safer way to produce power.

DID YOU KNOW? Philo Farnsworth received more than 300 **patents** during his life.

Pem stands next to a picture of Philo and holds his journal.

When Philo died in 1971, few people knew about him. Pem worked to change that. She wrote and spoke about his

inventions. Philo's TV never made him rich. But today, he is remembered for helping make television common.

Later TVs continued to use many parts Philo invented.

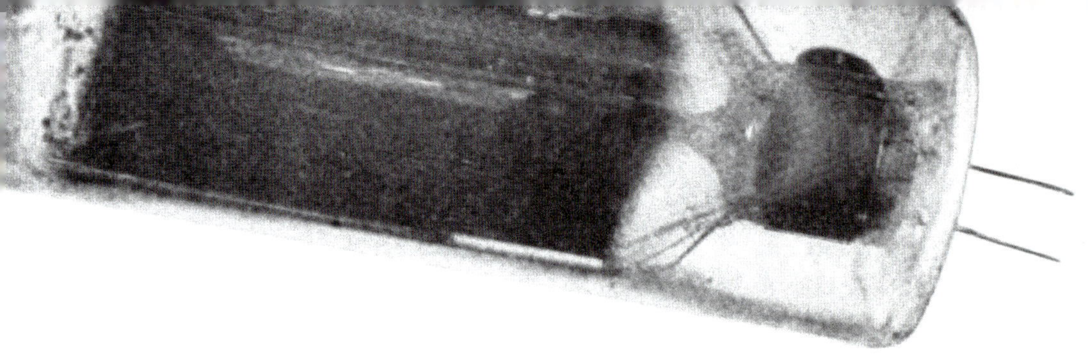

MAKING CONNECTIONS

TEXT-TO-SELF

Philo loved to read books and work on machines. Do you prefer to learn by reading about things or by trying them yourself?

TEXT-TO-TEXT

What other inventors have you read books about? What did Philo have in common with them? How was he different?

TEXT-TO-WORLD

Philo's invention helped solve problems with mechanical TVs. What is one problem facing technology today?

GLOSSARY

baby incubator – a tool doctors use to care for babies who are born too early.

broadcasting – the process of sending out signals that will be picked up by radios or TVs.

electron microscope – a tool that magnifies tiny parts of plants and animals so scientists can see them.

nuclear energy – energy released by combining or splitting tiny particles called atoms.

patent – rights given to an inventor for making a new object or tool.

receiver – the part of a TV that picks up signals and changes them into images.

INDEX

Scan this code* and others like it while you read, or visit the website below to make this book pop!

popbooksonline.com/philo-farnsworth

*Scanning QR codes requires a web-enabled smart device with a QR code reader app and a camera.